1 MONTH OF FREE READING

at

www.ForgottenBooks.com

By purchasing this book you are eligible for one month membership to ForgottenBooks.com, giving you unlimited access to our entire collection of over 1,000,000 titles via our web site and mobile apps.

To claim your free month visit: www.forgottenbooks.com/free171

* Offer is valid for 45 days from date of purchase. Terms and conditions apply.

ISBN 978-0-266-22163-0
PIBN 10000171

This book is a reproduction of an important historical work. Forgotten Books uses state-of-the-art technology to digitally reconstruct the work, preserving the original format whilst repairing imperfections present in the aged copy. In rare cases, an imperfection in the original, such as a blemish or missing page, may be replicated in our edition. We do, however, repair the vast majority of imperfections successfully; any imperfections that remain are intentionally left to preserve the state of such historical works.

Forgotten Books is a registered trademark of FB &c Ltd.
Copyright © 2018 FB &c Ltd.
FB &c Ltd, Dalton House, 60 Windsor Avenue, London, SW19 2RR.
Company number 08720141. Registered in England and Wales.

For support please visit www.forgottenbooks.com

CARPENTERS

INSTRUCTION MANUAL

No. 4

WAR DEPARTMENT
COMMITTEE ON EDUCATION AND SPECIAL TRAINING
WASHINGTON

CARPENTERS

INSTRUCTION MANUAL
No. 4

WAR DEPARTMENT
COMMITTEE ON EDUCATION AND SPECIAL TRAINING
WASHINGTON

WAR DEPARTMENT
COMMITTEE ON EDUCATION
AND SPECIAL TRAINING

INSTRUCTION MANUAL
No. 1

Ideals Back of Our Training Activities

There are two fundamentals that the Committee wishes to impress upon instructors:

1. **An efficient instructor must be accurately informed as to what jobs a carpenter, or a blacksmith, for instance, may be called upon to do in the Army.**

2. **Methods of instruction must be used which in the time available will best train men to do these jobs.**

To assist instructors in these two important respects, new course outlines are being prepared by the Committee on Education and Special Training. As soon as completed they will be issued to the co-operating schools.

These outlines are sufficiently definite to produce that degree of specialization required by the various Army Corps; yet they are also sufficiently flexible to allow for the broader development of resourcefulness and initiative which are essential to the men in the army. Results to be obtained have been set down rather than detailed methods of accomplishing results. Such details can be developed best in each institution through the ingenuity of the instructors in charge.

A promotion program is preferable to a rotating one. It is not desirable to follow rigidly a plan which provides that all students get a definite length of time on each of a number of specified subjects in rotation, though such a program is well suited to the most able men.

The work should be organized so that while the most able men will get the entire course of instruction, others will omit those subjects for which they have no special aptitude, giving full time and attention to those things in which they show promise of success. In general this means a course arranged not for the average but for the best student and then administered so that only those men who show or develop ability on one job will be promoted to the next more difficult task.

To illustrate: In the auto-mechanic squad only those men who have thoroughly mastered the details of the axle and wheel work will progress to engine work, and then again only those who show a good understanding of engine construction will take up further study of gas engine performance. Again only those who show special all-around ability in these things will proceed further with the study of ignition, timing and the more complicated details of operation.

In a given course all men need not start on the same job and all those promoted need not take up the same advanced work. There are several points in each course where men may be started. Obviously a man's first job should be of sufficient simplicity for him to proceed intelligently. All jobs should be reasonably sequential and should present real problems, the working out of which will equip the men with first hand knowledge of practical details of construction or operation and provide a background for the understanding of scientific principles. The important factor in shifting men from one job to the next is that the instructor maintain the idea and spirit of promotion as to both practical accomplishment and undersanding of fundamental principles. In short the educational program all the way through shall recognize native ability and previous experience and train men along the lines they are fitted to go and only as far as they can make real progress. At the end of the course each man will be rated, not on his average proficiency in all the work, but by listing the kinds of work he can do well.

In order to provide for the development of originality, initiative and real thinking power, and also to prevent learning by rule-of-thumb, the teaching should be almost entirely through jobs, questions, problems and guided discussions.

These outlines are not intended for the schoolroom but for the shop where they can be used for the guidance of the instructors and the men on the job. The frequent gathering of the men in small groups before a blackboard in the shop where live material is available for demonstration, discussion and conference is far preferable to the practice of formal lectures to large groups.

The accomplishment of a job is both the end to be attained and the means for instruction.

These ideals may not be new but are stated here as representing the composite ideas of all who are associated in the responsibility and conduct of this work.

Committee on Education and Special Training
C. R. DOOLEY,
Educational Director,
Vocational Instruction.

WAR DEPARTMENT
COMMITTEE ON EDUCATION AND SPECIAL TRAINING

COMMITTEE	ADVISORY BOARD

Brig. Gen. Robert I. Rees
  General Staff Corps
Col. John H. Wigmore
  Prov. Marshal General's Dept.
Lt. Col. Grenville Clark
  Adjutant General's Dept.
Major Wm. R. Orton
  War Plans Division
Major Ralph Barton Perry
  Executive Secretary
	James R. Angell
	Samuel P. Capen
	J. W. Dietz, Secretary
	Hugh Frayne
	Charles R. Mann, Chairman
	Raymond H. Pearson
	Herman Schneider

VOCATIONAL SECTION

C. R. Dooley, Educational Director

G. W. Hoke, Assistant Educational Director

EDITORIAL DIVISION

W. H. Timbie, Editor-in-Chief

J. A. Randall, Associate Editor

F. W. Boland........Carpenters and Sheet Metal Workers
H. D. Burghardt............................Machinists
F. A. Clark....Electricians and Mechanics (Negro Section)
S. L. Conner...............Surveyors and Topographers
F. H. Evans................Gunsmiths and Pipe Fitters
W. K. Hatt..........................Concrete Foremen
C. M. Jansky....................Telephone Electricians
R. A. Leavell...........................Auto Mechanics
W. H. Perry................................Electricians
E. M. Ranck..................................Farriers
W. R. Work........................Radio Electricians

WAR DEPARTMENT
COMMITTEE ON EDUCATION
AND SPECIAL TRAINING

CARPENTRY

Part I

GENERAL INFORMATION

Soldiers with certain kinds of technical skill constitute a valuable asset in all branches of the service, being as valuable in combatant units as in specialized Corps organizations. Our first and great task is to train our men as soldiers, for they must be part of a vast organization of soldiers. They have been sent to school in order that before entering upon overseas duty they may obtain as much skill and experience as possible in those trades which will greatly increase their effectiveness as soldiers and add to their usefulness to Corps to which they may later be assigned.

It is therefore highly desirable that schools training soldiers, reproduce at school, as far as possible, the army conditions, both as to methods of performing work and the equipment with which the jobs are done. It is equally important that the instruction material be selected directly from the jobs which the men are likely to be called upon to do in the army.

The following outline of a course for soldier carpenters has been prepared to assist instructors in reproducing the proper atmosphere and in selecting the most efficient methods and the most effective instruction material. It has been divided into four parts:

1. SPECIAL DUTIES of soldier carpenters.
2. THE TRAINING METHOD, which will in the time allowed, give the soldier the most experience in carpentry and the use of the necessary tools and developing his resourcefulness at the same time.
3. JOB SHEETS.
    A. Jobs which are to be performed by soldiers.
    B. Questions, the answering of which will test the soldiers' knowledge of the necessary procedure in doing the job and of the fundamental principles underlying this procedure.
4. SUPPLEMENTARY AIDS—Such as lecture notes, bibliography of manufacturer's instruction books and data sheets.

USING THE MANUALS

Parts One and Two on "Special Duties" and "Methods of Instruction" are primarily for the instructors' use. The remainder of the manual is to be used as Job Sheets which are to be handed to the soldier along with other aids at the time the job is assigned to him.

WAR DEPARTMENT
COMMITTEE ON EDUCATION
AND SPECIAL TRAINING

MANUAL No. 4
PART I
PAGE 2

The accomplishment of the job, rather than the ability to answer the test questions, is the end to be attained. A man may be able to answer practically all the questions and still not be qualified to do the work. The questions are for the purpose of calling the soldier's attention to the important features of the task and of ascertaining whether or not he has acquired the desired information from doing the job. In all the work, the ideals set forth in the instruction Manual No. 1 should be maintained.

WAR DEPARTMENT
COMMITTEE ON EDUCATION
AND SPECIAL TRAINING

CARPENTRY

Part II

TRAINING METHOD

GENERAL PROCEDURE:

In all cases the soldier should learn by doing. Accordingly, the work should consist of a maximum of shop or laboratory and outside work, in which he is assigned to specific tasks. For this purpose, the Job Sheets of Part Three have been prepared. In general, not more than three hours a week quiz work should be given, and this should deal principally with the tools, tool operations, materials and new principles of the jobs worked on. These quizzes or discussions may be divided under three heads as follows:

1. Materials, tools, care of tools and their uses.
2. A study of the job.
3. Estimating and sketching.

In addition to these discussions, short demonstrations, and conferences should be given to small groups before a blackboard set up in the shop.

No job should be attempted until a thorough study and general plan of procedure has been formed in the student's mind. This may be accomplished by making a sketch of the job, working up a complete bill of materials, and analyzing all of the tool operations.

This procedure will prevent the student from going at the job by poor methods, mistakes will be avoided, and material and time will be saved. It will compel him to use his own initiative on each task, which in the final analysis is the main end to be sought throughout his instruction period.

SCHEDULE

Accordingly a satisfactory method for training carpenters might consist of some such procedure as follows:

First: Have prepared by men in the detail, questionnaires showing each man's past experience. Trade test the soldier, if possible.

Second: Give a short talk on the absolute necessity of proper discipline in the shops, care of tools, equipment and materials, a general explanation of the course with emphasis put upon its importance.

Third: On basis of questionnaire and trade test, organize the class into groups as follows:

Group 1. Men with no experience.

Group 2. Men with little experience.

Group 3. Men with considerable experience.

Fourth: Assign each group to a job which the questionnaires show the men are qualified to undertake. The way in which they tackle the job or complete the task assigned them will show whether or not they have been assigned to the proper group.

Fifth: The men working at similar tasks should be called to a blackboard from time to time, as suggested above, and questioned concerning their work and the fundamental principles underlying the particular job at which they are working. Rate each man on each job performed. Sample record cards for this purpose can be obtained by applying to the committee, or a suitable form can be worked out at the school.

WORKING SCHEDULE

Prepare a schedule similar to the following:

Divide the allotted time, 8 weeks, into two equal periods of time. The first period, 4 weeks, to be used in following the general, or prescribed course. The second period, or remainder of time to be devoted to jobs of a more specialized character.

Section 1

First period of time

4 weeks—136 hrs.

- 24 hrs............ { Sketching / Questions / Demonstrations / Discussions / Mathematics }
- 112 hrs............ { Working out the prescribed jobs }

Section 2

Second period of time

4 week—136 hrs.

- 36 hrs............ { Sketching / Questions / Demonstrations / Discussions / Mathematics }
- 100 hrs............ { Working out prescribed jobs }

JOB SHEETS

A job sheet for this course in carpentry shows the following:

First: A working drawing of the job.

Second: A photograph of the completed or partly completed job.

Third: A statement of the job to be done.

Fourth: Questions on tools and tool operations involved in doing the job.

USE OF SHEETS

Since the job sheet shows a working drawing of the job, a statement of the job, and questions on the job, as well as on the various tools and tool operations, the men of a particular group, when assigned to a job, should be handed the job sheet.

The photograph, in most cases, is intended to serve a double purpose:

First: To assist the student in understanding and reading the working drawings by visualization. This will serve to teach the student the relationship between a working drawing and the finished job.

Second: It will show him by pictorial form how to go at the job. The questions are intended as an incentive to make a careful study of the job, of the tools and tool processes in doing the job, a knowledge of which is essential before attempting the task. Since students naturally take a great interest in questions, arguments and discussions as to their correct answers are likely to arise. These, in themselves, will stimulate the students' mental activities and tend to fix the information in their minds. It is much better for the instructor to act as a last court of appeal in settling these discussions, rather than as a source of information. The ideal instructor is not the one who imparts information to his class, but who stimulates and directs his classes so that by their own efforts the students gain the greatest amount of skill and information.

The instructor will find in each of the two sections of job sheets a larger number of jobs than any one man will have time to do. The extra number makes it possible to start each man on a job suited to his skill and to so select his jobs as to give him the maximum of training in the eight weeks available.

This course is divided into two sections. Section one consists of a number of jobs arranged with reference to the difficulty of the hand manipulation. Each job in the first section presents some project which cannot be done properly without the application of some new principle in carpentry as to tools, tool operation, or method of construction. The earlier jobs, are the ones that a rough carpenter will invariably be called upon to do first on new construction work. These jobs are all made with full size material. It is essential in that the soldier be made to work upon jobs having the atmosphere of reality.

It frequently happens in active operations that the most convenient source of material is nearby structures which may be dismantled. The material taken from such structures will be more valuable if considerable skill is used in dismantling or wrecking. Much of the material used in this course can be used a second and a third time. Just before the departure of each detachment, such finished work as is standing and for which there is no market should be dismantled by the soldiers and stacked up ready for the use of succeeding classes. A small amount of this training will not come amiss if care is taken not to dismantle articles at a time and in a manner to take away the atmosphere of real jobs and substitute the atmosphere of exercise work.

DUTIES

In practically all Corps of the Army there is a demand for rough carpenters. These mechanics should, first of all, have a thorough training in the use of tools, tool operations, should also, be trained to use their own initiative in construction work. A mastery of Section 1 of this course will give training in what may properly be called general carpentry work. Men with such training are needed in varying numbers in all Corps. Men trained in the part of the course outlined in Section 2 of this manual will be specially needed in the Engineer Corps, in the Field Artillery Corps, and in the Coast Artillery Corps.

Equipment: It is preferable to provide a special room for this carpenters course, rather than to attempt to use an equipped manual training shop, where small benches, having quick acting vises, are closely arranged in a small floor space. A ratio of free floor space unusual in school shops is required.

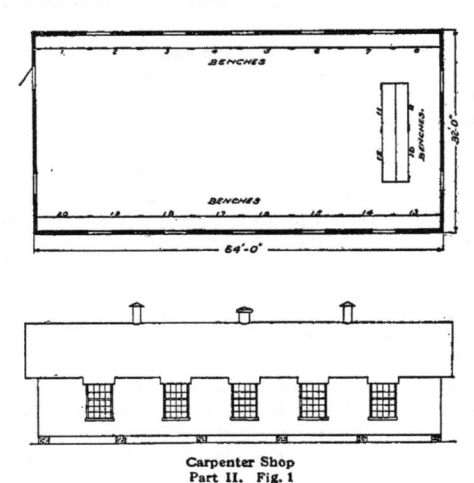

Carpenter Shop
Part II. Fig. 1

Wood working machine tools such as are found in most manual training shops may not be used in this course, as it is essential to teach carpentry for army purposes as a hand trade. The type of floor space recommended for the accommodation of a class of twenty men is supplied by a building measuring 32'x64'. A temporary structure similar to the one shown in following figure is entirely adequate and inexpensive.

SPECIFICATIONS AND SUGGESTIONS TO THE INSTRUCTOR

It is assumed in writing this manual, that the jobs be of such a character that the product produced will fill an actual need in the community. In some cases however, after the completion of the job it may be found that no available market can be found, in which case it may be desirable to tear down (dismantle) the article and use the material over again for the same or other jobs. With this in view, the following notes have been prepared on various job sheets as an aid to teachers to so specify sizes that he can attain the maximum use of his material in case it is necessary to tear down.

NOTES:
1. Material (Plank 12"x2"x8') used for first four jobs in this course, is to be used for a stair stringer in job No. 10. These first jobs are essentially processes in tool manipulation, accuracy and care in the handling of tools and materials which are essential requirements for a well trained mechanic. All of these points should be insisted upon by the instructor.
2. The 2"x4" used for saw horse should be rough (sawed) Hemlock or similar wood. Since it is to be used as a planing job it should not be used for a second saw horse, but may be used for bench legs or headers in house framing. The legs of saw horse may be used to advantage for nail box and bridging in building framing.
3. Material used for plumb rule may be used for this same job a second time by reducing length and width. It can be used later for saw horse legs, mitre box, nail box or for other purposes.
4. It would not be well to use material a second time in making a straight edge. This material can be used however in other construction work.
5. By decreasing size of rise, run, and width of stairs, the lumber used for these parts can be used a second time. These second stairs may be torn down and the materials used to make parts of other jobs, as mitre box, saw horse, work bench, nail box, etc. It is not likely that much material can be salvaged from stringers.
6. The initial carpenter's work bench when finished should not be less than 7'-10" long and 23" wide. It may be dismantled later and lumber used by other members of the class constructing a bench to be not less

than 7'-4" long and 22" wide. It may be dismantled a second and a third time and made not more than 4" less in length and 1" in width each time. Side and end boards should be made ½" less in width each time. Lumber should be used in constructing other jobs later. Nail washers should be used on all nails to facilitate their withdrawal in dismantling.

7. The jobs No. 13 and No. 14 headed "Building Framing" and "Roof Framing" are essentially group jobs; that is, these jobs should be given to a group of four students. Upon completing these jobs, and after credit has been given for the work, the building should be dismantled. By changing the pitch from one-third to three-eights or one-fourth, reducing height of building, (shortening studs) changing sizes of window and door openings, length of sill and joists, materials can be used over for the same jobs by several groups. It will, of course, be essential for the instructor to see that the roof plate has been framed up in advance. This roof plate should be mounted on saw horses and permitted to stand until all members of the class have finished cutting and fitting into place the three rafters which each is required to make. In planning this job for four students it was contemplated that each student in the group should lay out, cut and lay in place one hip, one common, and one jack rafter. It was further planned that all four students should work together to raise the ridge board and hip rafters.

As a second part of this job these same students should proceed to finish the roof frame referred to in Job No. 13.

CARPENTERS

INSTRUCTION MANUAL
No. 4

PART III. JOB SHEETS

WAR DEPARTMENT
COMMITTEE ON EDUCATION AND SPECIAL TRAINING
WASHINGTON

WAR DEPARTMENT
COMMITTEE ON EDUCATION
AND SPECIAL TRAINING

MANUAL No. 4
PART III

CARPENTERS

Part III

JOB SHEETS

DIRECTIONS FOR USING THE JOB SHEETS.

The first step in doing any carpentry job is to make suitable drawings or dimension the drawings on the job sheet and from them make a bill of materials.

Before starting on the assigned task study carefully the questions. Determine what things should be done, the method of doing them and the order in which they should be done. Questions on these points are always found at the beginning of the sheet.

The purpose of the remaining questions is to draw your attention to the important features and underlying principles with which you should become familiar through the doing of the task. Refer to the sheet from time to time as you progress in the work in order that you may be sure of observing the details which are important and may become familiar with the reasons for doing the job in a particular manner. It is only by thoroughly understanding the underlying principles of this job that you will be able to undertake intelligently similar jobs, depending upon the same principles. In this way, the knowledge gained from doing one task can be used to advantage in doing others.

SOURCES OF INFORMATION.

There are several available sources from which you may obtain the information to which the questions relate.

First: A study of the machines and tools which you are to use will furnish a fund of information that will be of value, not only in the job at hand, but in any similar job.

Second: Copies of Text Books, Hand Books, Data Sheets, Manufacturers' Catalogues and Instruction Books are available in the shop library. These should be consulted frequently.

Third: Some of the men working with you may have had previous experience which will enable them to help you. Do not hesitate to discuss the questions with the other men in the shop.

Fourth: The instructor may be appealed to for such information as you are unable to get in any other way. However, he will generally prefer to

WAR DEPARTMENT
COMMITTEE ON EDUCATION
AND SPECIAL TRAINING

tell you where the information can be found, and correct any misinformation which you may have obtained from other sources.

NOTE PARTICULARLY.

It is of utmost importance that you know the answer to every question on the Job Sheet and it is equally important that you obtain this information by your own exertions rather than have the instructor pass it out to you. Information obtained by experience is likely to be retained and become a part of your personal knowledge, while information obtained from the instructor is readily forgotten. Students should regard the instructor not so much as a source of information as a court of last appeal, who can settle discussions and rectify mistakes. It is much better to have the instructor judge as to the correctness of the information which you have obtained by your own efforts than to have him give you the information himself.

WAR DEPARTMENT
COMMITTEE ON EDUCATION
AND SPECIAL TRAINING

MANUAL No. 4
JOB No. 1
PAGE 1

CARPENTERS

Job No. 1

Prepare Stock for Stair Stringer. Saw to a Line*

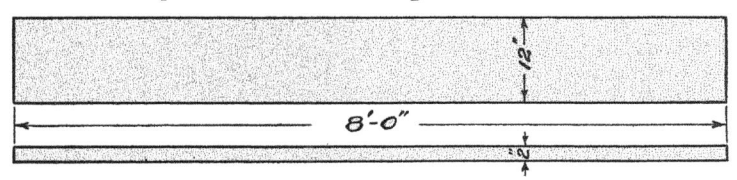

Working drawing
Job No. 1. Fig. 1

Square the ends of a 2' x12"x8' 0" plank by sawing 1" piece or pieces from the ends.

Operations: Lining (pencil and square), sawing, testing with steel square.

1. How does a plank differ from a board?
2. What are saw horses used for?
3. What is a steel square used for?
4. How should the square be held while marking a line from edge to edge? From face to face?
5. What is meant by squaring from edge to edge? From face to face?
6. What is a cross cut saw used for?

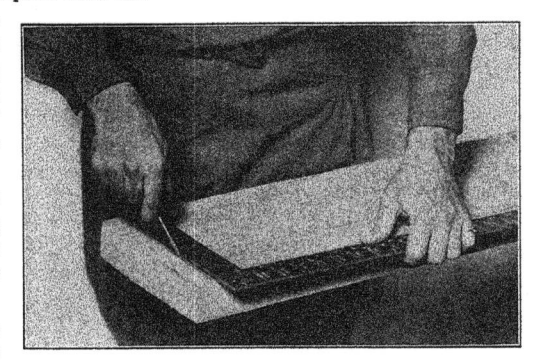

Use of the square
Job No. 1. Fig. 2

*Note: It may be necessary to make several cuts before attaining the desired results. (Not more than two feet of the plank to be used for this job). After each cut is made it should be carefully tested with the square.

Starting the saw cut
Job No. 1. Fig. 3

7. How does a cross cut saw differ from a rip saw?
8. What is meant by a 10 point cross cut saw?
9. How should the saw be held?
10. Tell just how to start the saw kerf.
11. How should the blade of saw be held so as to cut square with the surfaces of the plank?
12. If the saw leaves the line what should be done?
13. Should you saw on the line? If not, on which side of the line should you saw?
14. While sawing should you press upon the saw?
15. What should be done to prove that the saw blade is square with the surface of the plank?
16. How can you prevent the edge of the plank from splitting when finishing cut?
17. How can you prove the end of the plank has been sawed square?
18. How are saws designated as to size?
19. If you were going to purchase a cross cut saw for general use tell just what you would ask for.

Preventing the edge from splitting
Job No. 1. Fig. 4

WAR DEPARTMENT
COMMITTEE ON EDUCATION
AND SPECIAL TRAINING

MANUAL No. 4
JOB No. 2

CARPENTRY

Job No. 2

Prepare Stock for Stair Stringer. Rip to a Line.

Working drawing
Job No. 2. Fig. 1

Operations: Striking a chalk-line connecting two points, rip rawing, testing with try square.

Materials: Stock used for Job No. 1.

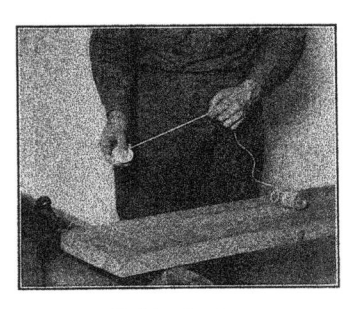

Chalking the line
Job No. 2. Fig. 2

Locate a point on surface of 2"x12"x 6' plank ¾" from edges at ends of plank. Connect points by striking a chalk line. Rip strip from plank being careful to follow chalk line and cut square to face of plank. It may be necessary to make two cuts before attaining desired results.

1. Is the term "a chalk line" always used to mean one thing? Explain.
2. How should line be held while being chalked? How should chalk be held?
3. Tell how to prevent line from cutting through chalk while chalking line.
4. What is a scratch awl? What is it used for?
5. How should the line be chalked?
6. Tell what is meant by snapping the chalk-line, and how it should be done.
7. What is a ripsaw? How does it differ from a cross cut saw?

Snapping the line
Job No. 2. Fig. 3

8. How should the saw be held while sawing?
9. What position should the operator take while sawing?
10. How should the saw kerf be started?
11. Should cut be made on the line, if not, on which side of line?
12. What should be done to keep saw from binding?
13. How should saw be cared for?
14. Why are the teeth in a rip saw filed differently from those in a cross cut saw? Explain.
15. Does the shape or the size of the teeth distinguish a rip from a cross cut saw?
16. How should edge of plank be tested after sawing?
17. If a plank is marked 9½" wide and it is ripped on the line will it be 9½" wide when finished? Explain.
18. If a plank 8" wide is to be cut down to 5" wide would you do it with a hatchet, rip saw or cross cut saw?
19. Tell something about the different methods of cutting planks from logs. Which is the best? Which is the most common method?
20. Name some of the defects due to irregularities in the growth of timber which render it unsuitable for carpentry.

Ripping to a line. Spreading the kerf
Job No. 2. Fig. 4

WAR DEPARTMENT
COMMITTEE ON EDUCATION
AND SPECIAL TRAINING

MANUAL No. 4
JOB No. 3

CARPENTRY

Job No. 3

Trim to a Line. Hand Ax or Hatchet. Prepare Stock for Stair Stringer.

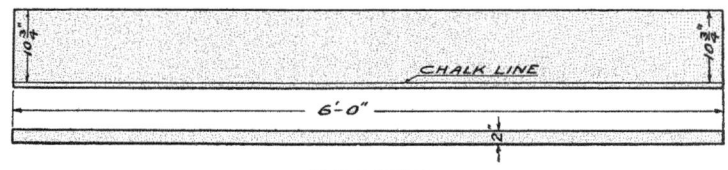

Working drawing
Job No. 3. Fig. 1.

Prepare stock for stair stringer. Trim to a line with hand ax or hatchet.

Operations: Measuring, striking a chalk line, scoring and paring.

Locate two points on surface and near ends of plank used in Job No. 2, 10¾" from sawed edge. Strike a chalk line connecting points. Cut or score edge from end to end and down to line with hand ax or hatchet. Begin at opposite end of plank and pare off chips with hand ax or hatchet, being careful not to cut below the line. If job is not satisfactory a second line may be struck 10" from edge and cutting tried a second time.

Paring
Job No. 3. Fig. 2

WAR DEPARTMENT
COMMITTEE ON EDUCATION
AND SPECIAL TRAINING

MANUAL No. 4
JOB No. 3
PAGE 2

1. How does a hand ax differ from a hatchet?
2. How should a hand ax or hatchet be held?
3. In measuring across plank how should rule be held?
4. What is meant by scoring? Why is it done? How is it done?
5. In scoring about how far apart should cuts be made? How deep should cuts be made?
6. What is meant by paring? How should it be done?
7. How does paring differ from scoring?
8. Should you try to get a finished surface with hand ax or hatchet?
9. Is the hatchet classed as a finishing tool?
10. Tell just how to test your results for accuracy.
11. Tell something about the cleavage properties of wood.
12. Tell something about the hardness of wood. When is a wood said to be hard? Name some of the hard woods. Some of the soft woods.
13. Is hemlock considered a brittle or flexible wood?
14. How many varieties of hemlock are there? Which variety is considered the best?

WAR DEPARTMENT
COMMITTEE ON EDUCATION
AND SPECIAL TRAINING

MANUAL No. 4
JOB No. 4
PAGE 1

CARPENTRY

Job No. 4

Plane an Edge Straight and Square. Prepare Stock for Stair Stringer. Joint an Edge With Jack Plane.

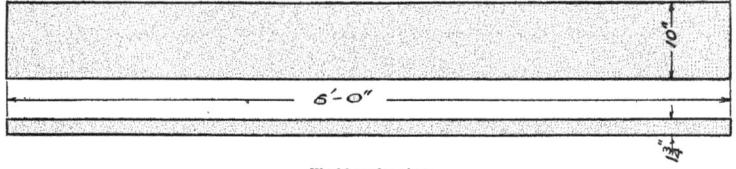

Working drawing
Job No. 4. Fig. 1

Prepare stock for stair stringer. Joint edge with jack plane.

Operations: Planing, use of try square, sighting. Joint edge with jack plane on plank used in Job No. 3 straight and square.

1. What is a plane used for?
   Tell how it should be held.
   Which is the toe of the plane? The heel?
   How should you start the plane? Finish?
   What is the plane iron? How is it held in place?
6. Tell how to adjust plane iron if it cuts more on one side than the other? If it cuts too deep?
7. How will you know when edge of plank has been planed square?
8. How will you know when edge is straight from end to end of plank?
9. What is meant by squaring?
10. What is meant by sighting along an edge?
11. Tell just how to sight along an edge.
12. How should work be held while being planed?
13. Tell some thing about pine—its general characteristics, etc.

Planing
Job No. 4. Fig. 2

WAR DEPARTMENT
COMMITTEE ON EDUCATION
AND SPECIAL TRAINING

MANUAL No. 4
JOB No. 5

CARPENTRY

Job No. 5

Make a Saw Horse Shown by the Accompanying Photographs and Drawings

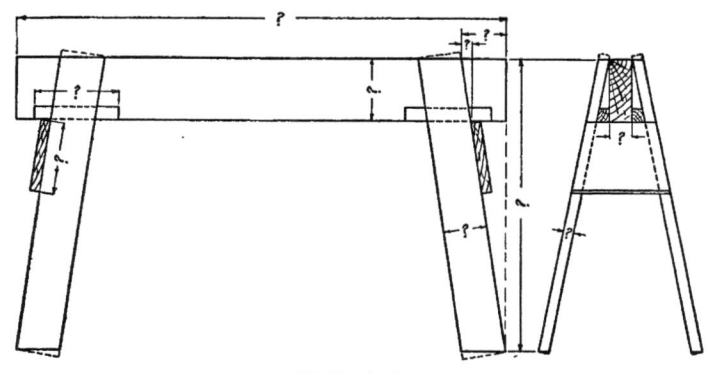

Working drawing
Job No. 5. Fig. 1

Operations: Laying out, sawing, planing, nailing, leveling up with rule and straight edge.

1. What is meant by a 2"x4'x3'—6" plank? How many board feet in this plank?
2. Name the several steps and tools used in squaring up the 2"x4"x3'—6" plank for saw horse.
3. What is a bench stop? For what used?
4. How should a hammer be held while driving nails?
5. What is meant by an 8 penny nail? A 16 penny nail? How many 8d nails come to a pound?
6. Tell just how you would withdraw a 16 penny nail.
7. Tell how you would lay off length of legs.
8. Should the legs of a saw horse be cut to the exact length and proper slant (bevel) before assembling? Explain.
9. What kind of wood will you use in making saw horse? Tell something about this wood.

WAR DEPARTMENT
COMMITTEE ON EDUCATION
AND SPECIAL TRAINING

MANUAL No. 4
JOB No. 5
PAGE 2

10. How many board feet of lumber will it take for this job?
11. Which saw (rip or cross cut) should be used for cutting on a slant?
12. Make up a complete bill of material for this job.
13. What is meant by a bill of material?
14. How should work be held while nailing?
15. Is there another way to make a saw horse than the one shown by drawing? If so make a sketch showing it. Tell of its advantages, if any, over the one shown by drawing.
16. Why are blocks placed under legs? How long should they be? How thick?
17. Tell just how to lay off for legs so all will have same slope. How much slope should they have?
18. Name various steps—their order in making saw horse.
19. How high should a carpenter's saw horse be?

Saw horse
Job No. 5. Fig. 2

WAR DEPARTMENT
COMMITTEE ON EDUCATION
AND SPECIAL TRAINING

MANUAL No. 4
JOB No. 6
PAGE 1

CARPENTERS

Job No. 6

Make the Nail Box Shown by Photograph and Drawings

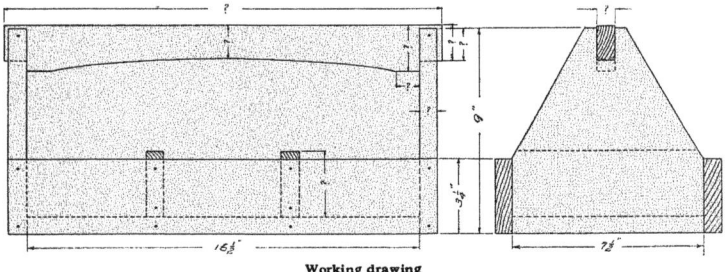

Working drawing
Job No. 6. Fig 1.

1. Make a sketch showing all necessary dimensions for nail box.
2. Make up an itemized bill of material giving length, width, thickness, and number of pieces of each size for this job.
3. What kind of wood would be best suited for this job? Why? Tell something about the structure and growth of trees, grains of wood, and defects due to irregularities in the growth of the tree.
4. What size and what kind of nails should be used? How many come to a pound?
5. If you know of a better way to fit handle to end pieces, show it on your sketch and submit to your instructor.
6. Which direction should the grain of wood run in end pieces? Why?
7. Would this size nail box be satisfactory for a general carpenter? Why?
8. Name the tools used in making this job.
9. Tell some of the various uses of a draw knife.
10. Tell just how the end pieces for this box should be laid out.

Nail box
Job No. 6. Fig. 2

11. Why is a cove used under tread? Should it be carried around under the end of the tread?
12. What is meant by a mitre cut? How should work be held in mitre box while being cut?
13. At 5½ cents per foot how much would the lumber be worth for these steps?

WAR DEPARTMENT
COMMITTEE ON EDUCATION
AND SPECIAL TRAINING

MANUAL No. 4
JOB No. 7

CARPENTRY

Job No. 7

Make a Plumb Rule Shown by Accompanying Photograph and Drawing

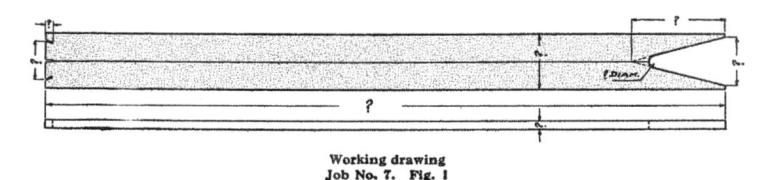

Working drawing
Job No. 7. Fig. 1

1. How is a plumb rule used? Describe some jobs requiring its use.
2. Name the various steps in their order in preparing stock for plumb rule.
3. What are the three small saw cuts for at top of plumb rule? How are they made? How far apart should they be? How deep should they be cut?
4. Why is notch cut out at lower end of plumb rule?
5. Why is it best to bore a hole as shown before cutting notch? What size bit should be used for this job?
6. Why should extreme care be exercised in making a plumb rule?
7. Tell why a center line is necessary. What tool should be used in drawing it?
8. If edges of rule and centre line are not parallel what will happen?
9. Name the tools used in making plumb rule.
10. What is a plumb bob? How does it work?
11. Tell how to test a plumb rule for accuracy.
12. If you find yours is not accurate how will you proceed to make it so?
13. What kind of lumber do you think will be best suited for this job? Why?
14. Small blocks of equal thickness are often nailed on one edge and near ends of plumb rule. Why?

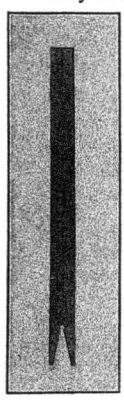

Plumb rule
Job No. 7. Fig: 2

WAR DEPARTMENT
COMMITTEE ON EDUCATION
AND SPECIAL TRAINING

MANUAL No. 4
JOB No. 8

CARPENTRY

Job No. 8

Make a Straight Edge Shown by the Accompanying Photograph and Drawing

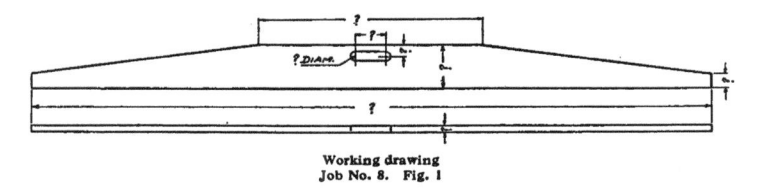

Working drawing
Job No. 8. Fig. 1

Operations: Laying out, sawing, planing, jointing, boring, chiseling, testing.

1. What use is there for a straight edge in carpenter work? Why not use a chalk line in all classes of work?
2. Should a chalk line or straight edge be used on short and accurate work?
3. Tell how to lay out handle on straight edge.
4. How should you proceed to cut out the handle of the straight edge?
5. What is a brace? How used?
6. What is an auger bit? How does it work?
7. What is meant by a number 13 auger bit? What size auger bit should be used for this job?
8. Tell just how the straight edge should be used to level up two joists 7 ft. apart.
9. Tell just how to test edge for straightness.
10. Tell just how to test a straight edge so that you are certain yours is true.
11. What is a level?
12. Tell how to test a level.
13. What is the difference between a paring chisel and a mortising chisel?
14. Tell what is meant by leveling with straight edge.
15. How will you get the two edges of the board parallel?
16. How does a jack plane differ from a smooth plane?
17. How can you be sure the holes are being bored perpendicularly through the board?

WAR DEPARTMENT
COMMITTEE ON EDUCATION
AND SPECIAL TRAINING

MANUAL No. 4
JOB No. 8
PAGE 2

18. How can you prevent splitting on opposite side of board from which you are boring?
19. If you find your straight edge is not true when tested how will you proceed to make it true?
20. Why is top edge of board tapered off as shown by photograph and drawing?

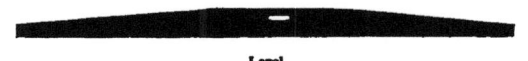

Level
Job No. 8. Fig. 2

WAR DEPARTMENT
COMMITTEE ON EDUCATION
AND SPECIAL TRAINING

MANUAL No. 4
JOB No. 9
PAGE 1

CARPENTRY

Job No. 9

Make the Mitre Box Shown by Accompanying Photograph and Drawing.

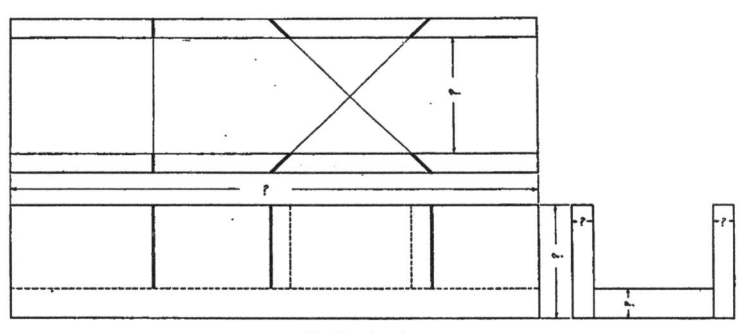

Working drawing
Job No. 9. Fig. 1

1. Why is a mitre box so named?
2. Is it necessary that a mitre box be straight and carefully constructed? Why?
3. How do you propose laying off the angles for the mitre cut.
4. Tell just how you would test your mitre box for accuracy?
5. Should it be necessary to plane a joint after it has been cut in mitre box?
6. What kind of a saw should be used to cut mitres?
7. How should mitre box be held while being used?
8. Name a number of places where mitre joints are used in house construction.
9. What size would you make a mitre box in order to do most of the work of a house carpenter?
10. Would you cut a mitre on a 12"x12" timber in a mitre box? If not tell how you would lay it out and make the cut.
11. What kind of lumber do you think would be best suited for a mitre box?
12. Tell something about the distinguishing features of the wood used for this job. What is meant by heart shake? Wind shake?

WAR DEPARTMENT
COMMITTEE ON EDUCATION
AND SPECIAL TRAINING

MANUAL No. 4
JOB No. 9
PAGE 2

13. If nails are used for this job how many and what size nails should be used? If screws are used what size and number should be used?
14. How many of the size nails you are going to use for this job come to the pound? How many screws come to a box? How are they designated as to size?
15. What is a gimlet bit? What size bit should be used for a No. 10 wood screw?
16. What is a countersink used for? How is it used?

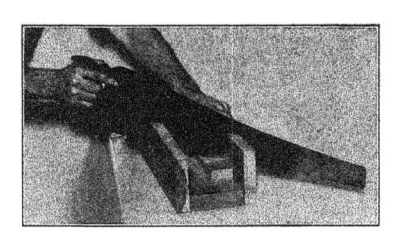

Mitre
Job. No 9. Fig. 2

WAR DEPARTMENT
COMMITTEE ON EDUCATION
AND SPECIAL TRAINING

MANUAL No. 4
JOB No. 10

CARPENTRY

Job No. 10

Lay Out and Cut the Porch Stair Stringers Shown by the Accompanying Photographs and Drawings.

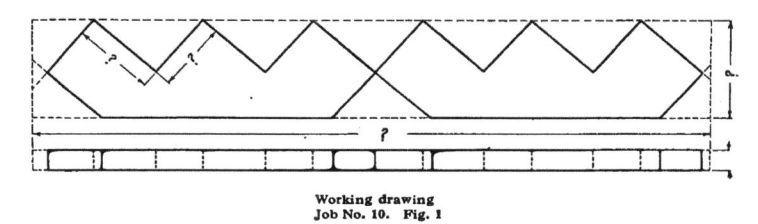

Working drawing
Job No. 10. Fig. 1

Use stock prepared in job 4.

The stringers are to be used for stairs to cover a 28¾" rise from ground level to porch level and a 30¾" run, with a total of four (4) risers to porch level.

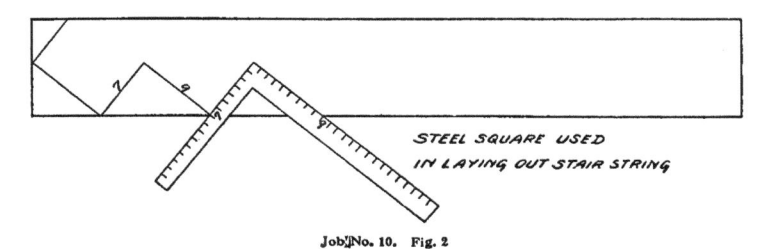

Job No. 10. Fig. 2

See photograph and drawing of stair Job No. 11. Also notes Part 4.

1. What is a stair stringer used for?
2. What is a good average rise and tread for a porch stair?
3. How many inches taken on the tongue and how many on the blade of square will give the cut in string for tread and riser?
4. Tell just how you determined width of tread and height of riser.
5. What is a good rule to follow in determining the relation of tread to riser?

WAR DEPARTMENT
COMMITTEE ON EDUCATION
AND SPECIAL TRAINING

MANUAL No. 4
JOB No. 10
PAGE 2

Sawing stair string
Job No. 10. Fig. 3

6. What is a framing square? Which part is the blade? The tongue?
7. Describe a pitch board and the manner of using it.
8. Tell how to make a pitch board after the tread and rise has been determined. Make one to be used for this job.
9. Are all the risers the same height in a string? If not which one is the lowest? How much lower is it than the others?
10. In cutting the treads and risers in a string should you cut right on the line or along side of the line? If the latter, along which side of the line?
11. Which saw should be used in cutting out a string?

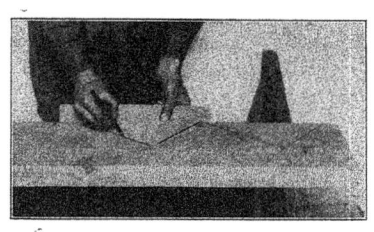

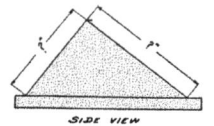

Pitch board in use
Job No. 10. Fig. 4

Job No. 10. Fig. 5

WAR DEPARTMENT
COMMITTEE ON EDUCATION
AND SPECIAL TRAINING

MANUAL No. 4
JOB No. 11

CARPENTRY

Job No. 11

Make the Porch or Basement Stairs Shown by Accompanying Photograph and Drawings.

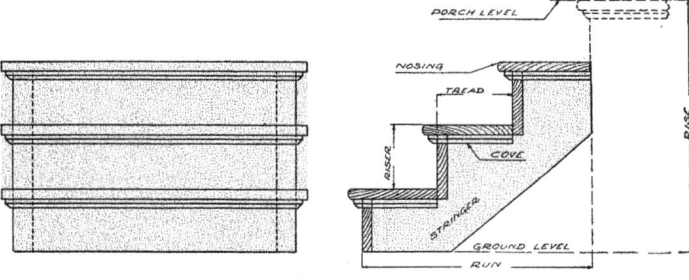

Working drawing
Job No. 11. Fig. 1

1. In planning a stair what is the first requirement?
2. What is meant by the riser? By the tread?
3. What is a good average amount in inches for the rise and tread for porch steps?
4. What is meant by steps being too steep? Too low?
5. What is meant by the total run? Total rise?
6. Are there more risers than treads? If so, how many?
7. In building stairs should the risers be nailed in place before the treads? If so, why?
8. What precautions should be taken in planing the end of a board?
9. What kind of nails should be used for this job if neat work is required?
10. Should the treads overhang at their ends? If so, how much?

Stairs
Job No. 11. Fig. 2

WAR DEPARTMENT
COMMITTEE ON EDUCATION
AND SPECIAL TRAINING

MANUAL No. 4
JOB No. 12
PAGE 1

CARPENTRY

Job No. 12

Make the Carpenter's Work Bench Shown by the Accompanying Photographs and Drawings.

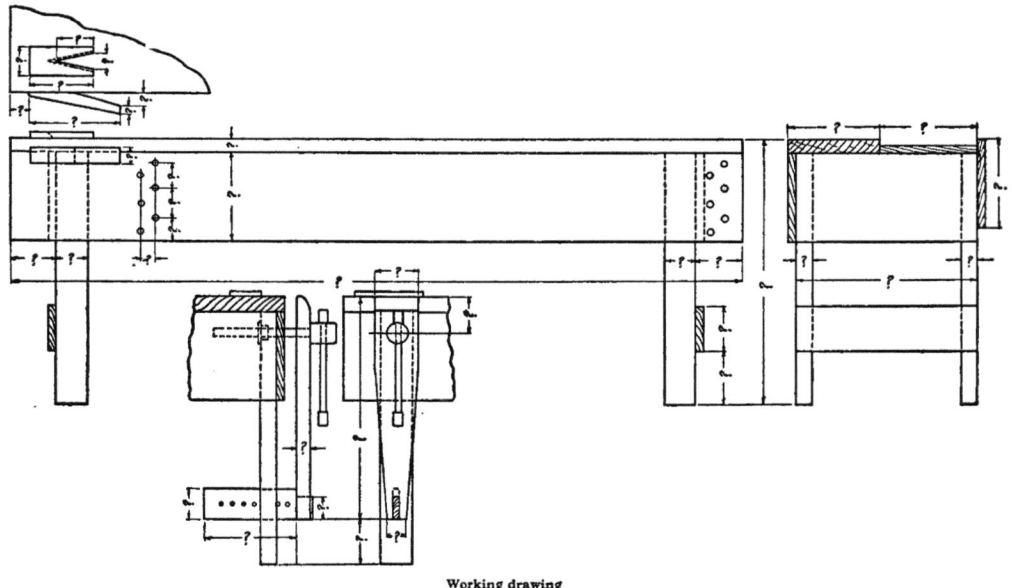

Working drawing
Job No. 12. Fig. 1

1. Tell what is meant by a working drawing. What should the working drawing show? Does it differ from a photograph? If so, in what way?
2. What is meant by front elevation? End elevation? Top view?
3. The drawing shows two types of vises. Which one is the better? The cheaper? When would it be necessary to use the small one shown on elevation and plan?
4. Why are top boards of different thicknesses used on bench? Are carpenter benches always made this way?

Carpenter's Work bench
Job No. 12. Fig. 2

5. Holes are bored through front board or apron. Why? Why are they staggered? What is meant by staggered holes?
6. These holes are to be ¾" in diameter. What No. bit will you use in boring them?
7. What precautions should be taken in boring through a board?
8. Make a free hand sketch of this bench. Give all dimensions and sizes of lumber to be used.
9. Make up a complete bill of materials for this job.
10. If you are going to use vise screw tell just how you will make out an order for it.
11. How high should a carpenter's bench be? How wide? Why are legs of bench placed as shown in drawing? Would bench be as rigid if legs were placed with narrow edge to front?
12. How wide should vise jaw be? How long should board be for vise jaw? If vise screw is 1⅛" in diameter what kind of a bit will you use in boring hole for screw?
13. If edges of top boards are crooked so they will not fit closely, how should they be prepared? What is meant by jointing two edges? What tools should be used? What is meant by a warped board?

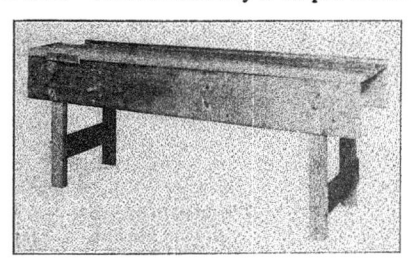

Rough work bench for carpenters
Job No. 12. Fig. 3

WAR DEPARTMENT
COMMITTEE ON EDUCATION
AND SPECIAL TRAINING

MANUAL No. 4
JOB No. 13
PAGE 1

CARPENTERS
Job No. 13

Frame the Corner of a Building

Dimension the drawing showing plan, front and side elevations.

Prepare a bill of material for corner of building shown by accompanying photographs and drawings. Frame the corner of building using the dimensioned drawings to work from. This is to be a group job. Four students to a group.

1. When we speak of framing a building, what is meant?
2. What is the difference in the framing for a full framed and a balloon framed building? Which type is the most common? Why?

Building framing
Job No 13. Fig. 1

3. What part of the framework is the sill? Make a sketch showing a sill for a balloon framed building, also show method for fastening and supporting joists.
4. What size plank will you use for joists for this job? How many board feet in a 2"x10"x6' plank? How far should joists be spaced apart? Tell how to use the level board in leveling up joists.
5. What is meant by floor bridging? Tell how to get the proper bevel for cutting bridging. How are they usually cut? What size stock is usually used? About how far apart are they usually spaced? If a sufficient number of bridging, properly cut and nailed in place,

Photograph Showing Door and Window Frames in Place
Job No. 13. Fig. 2

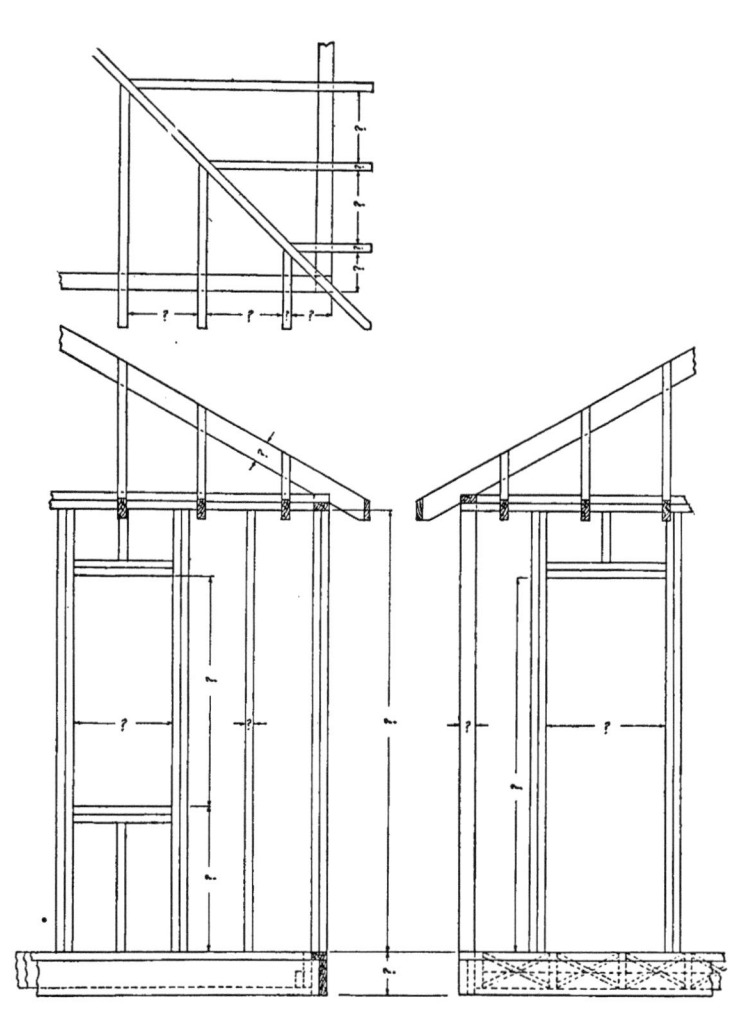

WAR DEPARTMENT
COMMITTEE ON EDUCATION
AND SPECIAL TRAINING

MANUAL No. 4
JOB No. 13
PAGE 2

Working drawing of building frame
Job No. 13.   Fig. 3

is not used, what is likely to happen? How many nails should be used to fasten bridging?
6. What size stock is usually used for studs? How many board feet in a stud measuring 2"x4"x8' 0"? How far apart should studs be spaced? Why? Why are double studs used for corner posts? What is meant by plumbing a corner post? How is it done? What tool is used? Why are double studs usually used at openings for doors and windows? What are the horizontal pieces called that are placed over door and window openings? How are they held in place? What are the studding called that are placed below window openings?
7. Make sketch showing how corners of plate are jointed. What size stock should be used for plate?
8. If door is to be 2'-6" wide and 6'-10" high, how large should the opening be for door frame?
9. If window is to be of double hung type and glass is to be 24"x30", how large should opening for window frame be?
10. Tell how you can prove that the door and window openings are perfectly square.
11. How high should the top of the window stool be above joists?

Plumbing corner post
Job No. 13. Fig. 4

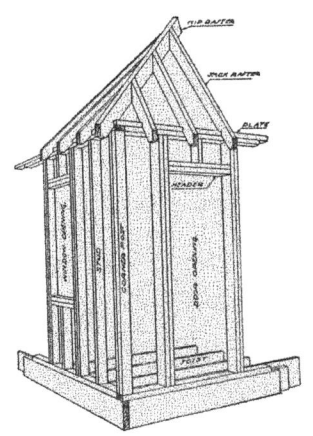

Perspective of building framing
Job No. 13. Fig. 5

WAR DEPARTMENT
COMMITTEE ON EDUCATION
AND SPECIAL TRAINING

MANUAL No. 4
JOB No. 14
PAGE 1

CARPENTERS

Job No. 14

Frame a Roof

Put the dimensions upon the drawing in the following figure. Compute the lengths of the hip rafters, the common rafters, and the jack rafters for a roof to cover a building measuring nine feet wide by twelve feet long and having a one-third pitch.

Note: See "roof framing" in Part IV, page 12.

1. What is meant by a one-half pitch roof? A one-third pitch roof? A one-fourth pitch roof?
2. How will you proceed to find the length of a common rafter?
3. Should any allowance be made for thickness of ridge board in figuring length of common rafters? If so, how much?
4. How do you find the length of a hip rafter?
5. How do you find the length of a jack rafter?
6. What do the terms "Rise" and "Run" indicate as applied to roof framing? Are these terms used in any other connection in building construction?
7. If the run for a certain roof is 12 feet and the roof is to be one-third pitch, what is the rise?
8. What does the term "heel cut" mean as applied to roof framing?
9. Make sketch showing how to lay off heel cut, using the framing square, on common rafters, hip rafters and jack rafters.

Note: See "roof framing," Part IV, page 12.

10. In what different positions is the square placed in laying off heel cut on a common rafter, jack rafter and hip rafter?
11. Make a sketch showing how to lay off the plumb cut on a common rafter, jack rafter and hip rafter.
12. In what different positions is the square placed in laying off plumb cut on a common rafter, jack rafter and hip rafter?
13. Make a sketch showing how to lay off the side cut on a hip rafter. Is there a side cut on a common rafter? On a jack rafter?
14. What do the terms "end cut," "side cut," "plumb cut," and "heel cut," mean as applied to roof framing?

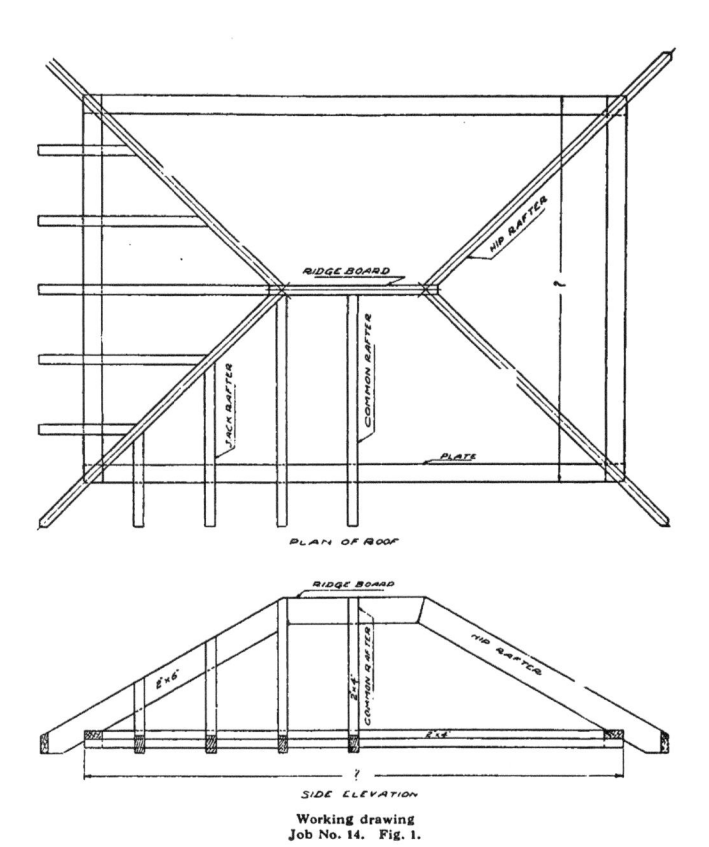

WAR DEPARTMENT
COMMITTEE ON EDUCATION
AND SPECIAL TRAINING

MANUAL No. 4
JOB No. 14
PAGE 2

Working drawing
Job No. 14. Fig. 1.

WAR DEPARTMENT
COMMITTEE ON EDUCATION
AND SPECIAL TRAINING

MANUAL No. 4
JOB No. 14
PAGE 3

15. Make a layout, 1½" to the foot (similar to Fig. 1, Part IV, roof framing) for a ½ pitch roof. Building to measure 9' 6" x 12' 6". Find length of rafters.
16. Is there any method by which length of common rafter can be found without making a layout? (Similar to Fig. 1, Part IV, roof framing).
17. Tell how you would find the length of a common rafter by scaling.
18. What are the lengths of the common and hip rafters of a one-third pitch roof, width of building across top plate being 18 feet? How long should the ridge board be if the building is 36 feet long?

Photograph of Hip Roof
Job No. 14. Fig. 2

Scaling
Job No. 14. Fig. 3

| WAR DEPARTMENT | MANUAL No. 4 |
| COMMITTEE ON EDUCATION AND SPECIAL TRAINING | JOB No. 15 PAGE 1 |

CARPENTERS

Job No. 15

Make the Window Frame Shown by the Accompanying Photograph and Drawings

Make a dimensioned sketch of this window frame, also a bill of material.

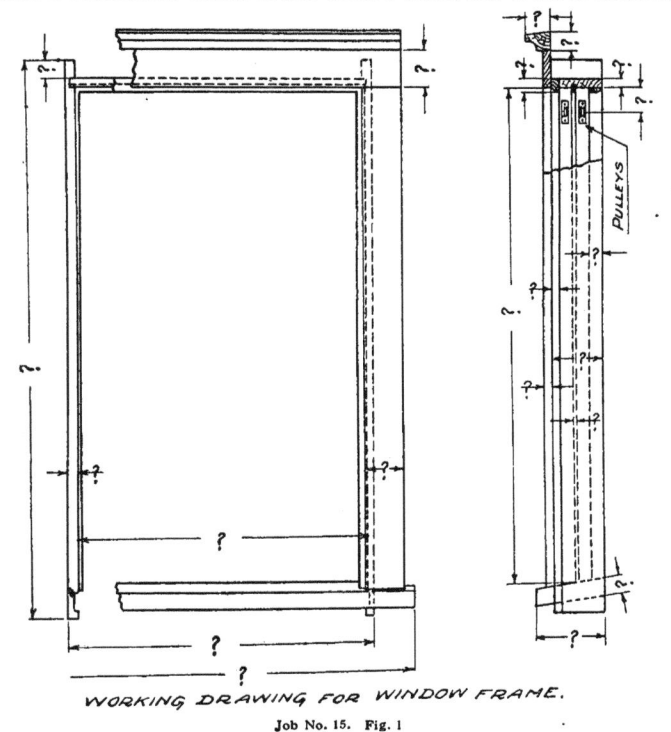

WORKING DRAWING FOR WINDOW FRAME.
Job No. 15. Fig. 1

1. What should be the width of a window frame between jambs or pulley stiles to accommodate a sash for glass 20" wide?

WAR DEPARTMENT
COMMITTEE ON EDUCATION
AND SPECIAL TRAINING

MANUAL No. 4
JOB No. 15
PAGE 2

2. What should be the length of frame from top of sill to under side of jamb for a double hung window, glass to be 28" long?
3. How much pitch should the sill be given?
4. Jambs or pulley stiles are grooved for parting strip. Why? What tool should be used for this?
5. How much space must be allowed between jambs and studs for weights? What is this space called?
6. Are pulleys placed in jambs before or after the jambs are assembled?
7. How can you prove that window frame is square?
8. What kind and size nails should be used for this job?
9. After job has been properly dressed with smooth plane it should be sanded. What number sand paper should be used? Which is the finer No. 1 or No. 1½ sand paper?
10. When we speak of housed joints, what is meant?
11. Housed joints are used on this job. How deep should they be cut? What tools are used in making a housed joint?

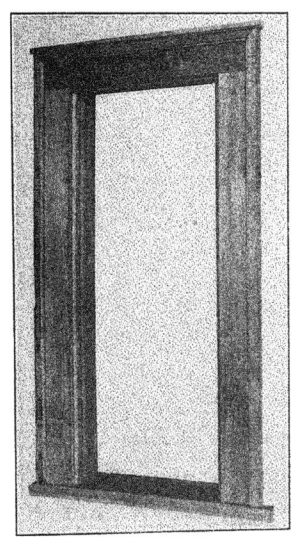

Photograph of Window Frame
Job No. 15. Fig. 2

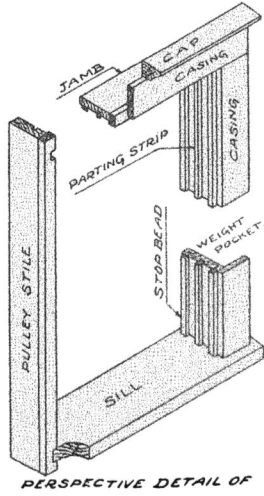

PERSPECTIVE DETAIL OF WINDOW FRAME.
Job No. 15. Fig. 3

WAR DEPARTMENT
COMMITTEE ON EDUCATION
AND SPECIAL TRAINING

MANUAL No. 4
JOB No. 16
PAGE 1

CARPENTERS

Job No. 16

Make the Door Frame Shown by the Accompanying Photograph and Drawings

Make a dimensioned sketch of this door frame, also a bill of material.

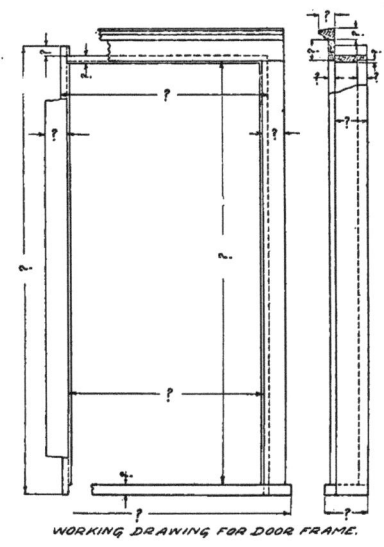

WORKING DRAWING FOR DOOR FRAME.
Job No. 16. Fig. 1

1. What is meant by a rabbeted jamb? What tool should be used for rabbeting jamb?
2. Should ends of sill be housed into jambs; if so, how deep? Should sill be nailed to jamb or jamb to sill? What size nails should be used?
3. What should be the pitch or fall of door sill?
4. Tell just how to assemble parts of frame after they have been prepared.

WAR DEPARTMENT
COMMITTEE ON EDUCATION
AND SPECIAL TRAINING

MANUAL No. 4
JOB No. 16
PAGE 2

Should side casings be cut to length and nailed or should they be first nailed and then cut to length?
5. Name several kinds of lumber suitable for this type of work? Why? Tell something about the mechanical properties of this wood.

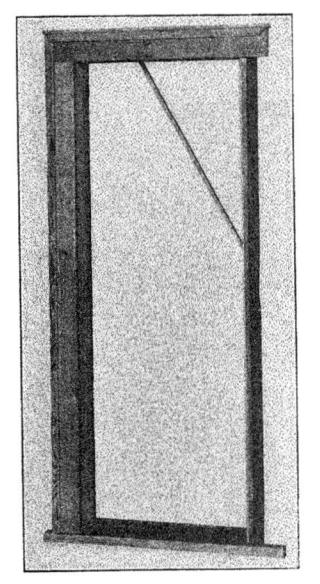

Photograph of Door Frame
Job No. 15. Fig. 2

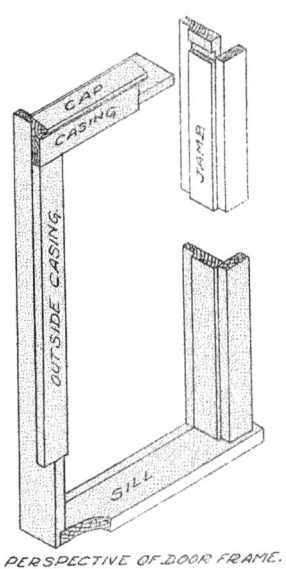

PERSPECTIVE OF DOOR FRAME.
Job No. 16. Fig. 3

CARPENTERS

INSTRUCTION MANUAL
No. 4

PART IV. SUPPLEMENTARY INFORMATION

WAR DEPARTMENT
COMMITTEE ON EDUCATION AND SPECIAL TRAINING
WASHINGTON

WAR DEPARTMENT
COMMITTEE ON EDUCATION
AND SPECIAL TRAINING

CARPENTERS

Part IV

DESCRIPTION OF THE CONSTRUCTION AND USE OF THE VARIOUS TOOLS USED IN THIS COURSE

1. Awls, scratch; metal point, with wood handle; used as a pencil in laying off work on rough lumber.
2. Bits, auger; used in a brace for boring holes in timber.
3. Bits, expansion; used for boring larger holes than can be cut with the bit augers.
4. Bits, screw driver; used with brace.
5. Braces, ratchet; used for rotating the auger, screw driver, and expansion bits.
6. Chisels, cold; used for cutting metal.
7. Chisels, framing, handled; used with the carpenter's mallet for a variety of purposes, but largely for cutting mortises.
8. Dividers, wing; used in laying off work.
9. Drawknife; used for shaping timbers.
10. File, saw, taper; used for sharpening saws.
11. Hammer, claw; used for driving and pulling nails.
12. Hatchet; used for shaping timbers and for driving nails.
13. Level, carpenter's, 24-inch; used for leveling and plumbing.
14. Mallet, carpenter's; used for striking framing chisels and for other work that would be bruised were a hammer to be used.
15. Oilstone; for sharpening tools.
16. Jack plane; used for smoothing pieces of timber.
17. Plumb bob; 6-ounce; used for determining a vertical line.
18. Rules, 2-foot, 4-fold; used for laying off work.
19. Saw, cross cut, hand; used for cutting timber across the grain.
20. Saw, rip, hand; used for cutting timber along the grain.
21. Saw set; used for setting saws (that is, bending the teeth in alternate directions, so as to get the width of cut desired).
22. Screw driver; used for driving screws.
23. Steel square, carpenter's; used for a large variety of purposes when laying off work.

24. Try square; used largely for squaring timber.
25. Tape, metallic, 50-foot; used for laying off large work.
26. T bevel; used for laying off bevels after the piece has been squared.
27. Chalk, carpenter's; used for laying off work.
28. Chalk line, 40-foot; used for laying off work.
29. Pencils, carpenter's; used for laying off work.

LUMBER

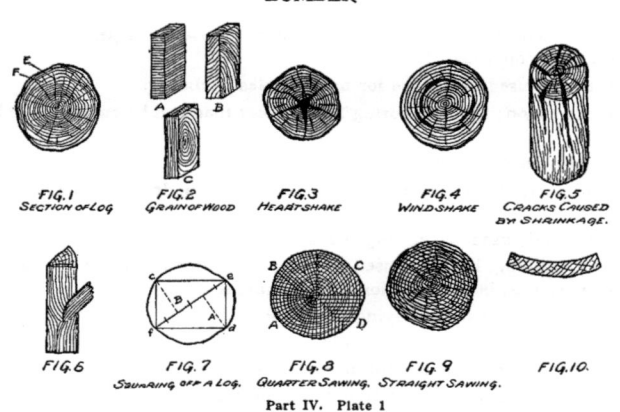

FIG. 1 SECTION OF LOG
FIG. 2 GRAIN OF WOOD
FIG. 3 HEARTSHAKE
FIG. 4 WINDSHAKE
FIG. 5 CRACKS CAUSED BY SHRINKAGE

FIG. 6
FIG. 7 SQUARING OFF A LOG.
FIG. 8 QUARTER SAWING.
FIG. 9 STRAIGHT SAWING.
FIG. 10.

Part IV. Plate 1

1. CLASSIFICATION OF TREES.

The trees from which most of our lumber is secured are of two kinds—the broad leaved, such as the oaks, poplar, and maples, and the coniferous or "needle leaved," such as the pines, cedars, etc.

2. STRUCTURE AND GROWTH OF TREES.

"If a cross section of a full-grown pine tree is carefully examined, at the center will be found a small pith; then a great number of concentric rings, varying in width and spacing; and finally an envelope of bark. The rings are alternately light and dark, one light and one dark ring representing a year's growth; the light wood is spring growth and is comparatively soft and weak; the dark ring is summer growth and is dense and strong. The strength of the wood may therefore be measured by the ratio of summer to spring wood in a unit of volume. In a cross section, the rings decrease in thickness from the center to the bark; hence the strongest timber will come from the lower part of the tree, midway between the pith and the bark—

"The sapwood is a zone of light, weak wood, 30 or more rings wide, next to the bark; the outer portion of it is the growing part of the tree. The heartwood is the inner and darker portion of the sector and has had no part in the growth of the tree; it is much stronger and denser than the sapwood. Heartwood results from the gradual change of sapwood due to the infiltration of chemical substances from the sap. The proportion of heartwood depends upon the age of the tree, forming about 60 per cent of an old, long-leaf pine."

It follows from this that heartwood is best, and that if obtainable at a reasonable price it should be used for all work where lasting qualities are important.

In addition to the annual rings, an examination of a cross section of any log reveals other lines radiating from its center. These are known as the "medullary" rays. Usually they do not extend to the bark, but alternate with others which start at the bark and run inward toward the center, but are lost before they reach the pith. This is shown by E and F of figure 1, plate 1.

3. GRAINS OF WOOD.

A piece of wood is said to be fine grained, coarse grained, cross grained, or straight grained. It is fine grained when the rings are relatively narrow, and coarse grained when they are wide. Fine grained woods will take a higher polish than coarse grained woods. When the fibers are straight and parallel to the direction of the trunk, the wood is said to be straight grained, but if they are twisted or otherwise distorted, the wood is said to be cross-grained. On plate 1, figure 2, A, is cross grained; B, partially cross grained; and C, straight grained.

4. DEFECTS.

The defects due to irregularities in the growth of the tree, which render timber unsuitable for the carpenter, are heart shake, wind shake, star shake, and knots. Other defects, due to deterioration of the timber both before and after it has been placed on a structure, are dry and wet-rot. Dry-rot is caused by a fungus growth and takes place most readily when the timber is so placed in a structure that it is alternately wet and dry. If it could be kept perfectly dry, or, on the other hand, constantly under water, it would last indefinitely. For this reason, piles should be cut off below the water level. Dry-rot takes place most rapidly when the wood is also confined, such as being buried in a brick wall, so that the gases of disintegration can not escape. This can be prevented by ventilation or by introducing certain salts of mercury and other metals into the wood.

Wet-rot is a form of decay which takes place in the growing tree. It is caused by the tree becoming saturated with water, and may be communicated from one piece of timber to another by contact.

Plate 1, figure 3, shows what is known as a heart shake. It is caused by the formation through decay of a small cavity in the heart of the tree which is followed by the formation of radial cracks.

Plate 1, figure 4, shows a wind shake which is caused by the separation of the annual rings so that an annular crack is formed in the body of the tree. Such a crack may extend for a considerable distance in the direction of the length of the tree. This defect is said to be caused by the alternate expansion and contraction of the sap wood and the wrenching to which a tree is subjected during high winds.

A star shake is very much like a heart shake, except that the cracks extend across the center of the trunk but without the appearance of decay at that point.

Warping of timber is the result of the evaporation of part of the water held in the cellular wall of the wood in its natural state, and the consequent shrinkage of the piece. If the timber were uniform in structure throughout the shrinkage would be the same in all parts and there could be no warping. As already explained, however, wood is made up of a large number of layers of different thicknesses in different parts of the log so that one layer may shrink more than another in drying. Because of the intimate connection between these layers one layer can not shrink or swell without changing the form of those adjacent. The timber as a whole must therefore adjust itself to the new conditions, and warping results. The only way to prevent it is to permit the log to dry out or "season" before sawing. After it is once thoroughly seasoned it will not warp unless it is allowed to absorb more moisture. Figure 5 of Plate 1 shows cracks caused by uneven shrinkage.

Knots are common in all timbers. They form at the junction of the main tree trunk and branches. At such points the fibers in the main trunk are turned aside and follow the branch as shown in figure 6, Plate 1. Often-times a branch will break off close to the trunk, and if the tree is still growing the end of the branch will be buried in the trunk. Meanwhile the branch dies and a knot is formed. Since the dead wood has no connection with the live wood around it, it will work loose in time and drop out when the tree is sawed into lumber. A knot as long as it remains in place does not seriously impair a timber subjected to a compressive stress, but it greatly weakens a piece subjected to tension.

5. SAWING OF LOGS.

    a. "Squaring" a Log.

If a log is to be "squared" so as to form a single timber a good rule to follow for determining the dimensions of the largest piece which can be cut from it is as follows: Divide the diameter into three equal parts and erect perpendiculars at these points as shown at A and B in figure 7 of Plate 1. The points c and d in which these perpendiculars cut the circumference, together with the points where the assumed diameter cuts the circumference will give the four corners of the timber. This is the largest and best timber that can be cut from the log.

b. Sawing Planks.

Plate 1, figure 8, shows several different methods of cutting planks from logs. In the first place a log is usually divided into quarters and the planks are cut as shown. The method shown at A is the best and is called quarter sawing. All the planks are cut radially from the center, so that the liability of splitting or warping is much reduced. A fairly good method is that shown at B. Method C, which is the most common, gives fairly good results, although the center plank is the only radially cut plank. Planks may be simply sliced from the log as shown in figure 9. This is the poorest method of all, as the natural tendency of the plank to shrink will cause it to curve as shown in figure 10. It is practically impossible to flatten such a plank.

6. PROPERTIES OF VARIOUS TIMBERS SUITABLE FOR CARPENTRY.

a. Coniferous—Evergreens.

(1) White Cedars.—There are five different kinds of white cedar in the United States, of which four are species of white cedar proper while the fifth is known as the canoe cedar. Cedar is light and soft, possesses considerable stiffness and a fine texture. "White" cedar is of a grayish brown color, the sapwood being lighter than the heartwood. It seasons quickly, is durable, and does not shrink nor check seriously. Its principal use is for shingles, posts, and railroad ties. The trees are usually found scattered amongst other kinds, though they occasionally form quite considerable forests. They are found all through the northern part of this country and along the Pacific coast. Most of the trees are of medium size while others are very large, especially the canoe cedar of the Northwest.

(2) The red cedars are similar to the others but have a somewhat finer texture. There are two varieties, the red cedar proper and the redwood. The former is found principally in the Southern States and the latter only in California.

(3) Cypress—This occurs only in the southern part of the country where it grows in swamps or along low river banks. There are a great many varieties, of which the "Gulf cypress" is the best. The timber is light, straight grained, and soft, and is admirable for shingles, siding, water tables, sills, or gutters, since it has great resistance to warping and the effect of dampness.

(4) Hemlock—There are two varieties of hemlock, one found in the

Northern States from Maine to Minnesota and in the Alleghenies, southward to Georgia and Alabama, and the other in the West from Washington to California and eastward to Montana. The eastern tree is smaller than the western and its wood is lighter and softer and generally inferior. It is of a light, reddish-gray color, fairly durable, but shrinks and checks badly, is rough, brittle, and usually cross-grained.

(5) Spruce—There are three kinds of spruce—white, black, and red, of which the former is most commonly seen in the market. The wood is light and soft, is fairly strong, and is of a whitish color. The trees are small, so that the lumber can only be obtained in small size. It is quite satisfactory for light framing.

White spruce is found scattered throughout the Northern States.

Black spruce is found in Canada and in some of the Northern States. It is distinguished from the other varieties only by its leaves and bark.

Red spruce, sometimes known as Newfoundland red pine, is found in the northeastern part of North America. Lumber cut from it is similar to that cut from black spruce.

(6) Pines—The distinguishing features of the pines are their great height, strength, and freedom from many branches. For this reason longer and larger pieces of lumber can be obtained from them. Two distinct classes of pine used in building work, the soft and hard pine, are found in great abundance scattered throughout the entire United States. The softer varieties are used for outside finishing of all kinds, and the harder for heavy framing and floors. There are two kinds of soft pine, the white pine and the sugar pine, the latter being a western tree found in Oregon and California, while the former is found in the Northern States from Maine to Minnesota. A smaller species of white pine is also found along the Rocky Mountains from Montana to New Mexico.

There are ten different varieties of hard pine, of which only five are of practical importance to the carpenter. These are the long leaf southern pine, the short leaf southern pine, the yellow pine, the loblolly pine, and the Norway pine. The long leaf pine, also known as the Georgia pine and the long straw pine, is a large tree which is found from North Carolina to Texas. It yields very hard strong timbers which can be obtained in large sizes.

The loblolly pine is also a large tree. It has more sapwood than the long leaf pine, but is coarser, lighter, and softer. It is the common lumber pine encountered from Virginia to South Carolina. It is also found in Texas and Arkansas. In some places it is known by the name of slash pine, old field pine, rosemary pine, sap pine, short straw pine, or Texas pine. The short leaf pine is much like the loblolly pine and is the chief lumber tree of Missouri and Arkansas. It is also found in North Carolina and Texas.

The Norway pine is a northern tree found in Canada and our Northern States. It never forms forests, but is found scattered amongst other trees, but sometimes in small groves. The wood is fine grained and white in color. It consists largely of sapwood, so that it is not very durable.

Great care is necessary when ordering pine lumber to make sure that the purchaser and seller have the same wood in mind when they apply a particular name to it. White pine, soft pine, and pumpkin pine are terms used in the Eastern States for the timber taken from the white pine tree, while on the Pacific coast the same terms refer to the timber of sugar pine.. The name yellow pine when used in the East is generally applied to the pitch or southern pines, but in the West it refers to the bull pine. Georgia pine or long leaf pine is a term applied to the Southern hard pine which grows in the coast region from North Carolina to Texas and which furnishes the strongest pine lumber on the market. Pitch pine may refer to any of the southern pines or to the pitch pine proper, which is found along the coast from New York to Georgia and in the mountains of Kentucky.

b. Broad Leaved Trees.

(1) Ash—This is a wood frequently employed for interior finish. It shrinks moderately, seasons with little injury, and will take a good polish. The trees grow rapidly, but only to a medium height. They do not, as a rule, grow together in large numbers. Of the six different species found in the United States, only two, the white and black ash are used extensively by carpenters. The first is more common along the basin of the Ohio River, but also occurs from Maine to Minnesota and in Texas. The black ash (sometimes known as hoof or ground ash) is also found from Maine to Minnesota and southward to Virginia and Arkansas. There is very little difference between the two species.

(2) Beech—This is also used to some extent for inside finish. It is heavy, hard, and strong, but of coarse texture. Its color runs from white to brown. It shrinks and checks during the process of drying and is not durable when placed in contact with the ground. It works easily, stands well, and takes a good polish.

(3) Birch—This is a very handsome wood, brown in color but with a satin luster. It takes a good polish, works easily, and does not warp. It is not, however, durable in exposed positions. It is much used to imitate cherry and mahogany, as its grain is very similar to these woods. The trees are of medium size and are found throughout the eastern part of the United States.

(4) Butternut—This is light, soft, and weak. Its color is light brown. The trees, which are of medium size, are found in the Eastern States from Maine to Georgia.

(5) Cherry—The wood is heavy, hard and strong and of a fine texture.

The heart wood is of a reddish-brown color, while the sap wood is yellowish white. It takes a good polish, works easily, and stands well. It shrinks considerably during drying. Most cherry lumber is cut from the wild black cherry tree, which is of medium size and found scattered amongst other broad leaved trees along the western slope of the Alleghenies and as far west as Texas.

(6) Chestnut—Chestnut timber is used in cabinet work and interior finish, and sometimes for heavy construction. It is light, fairly soft, but not strong. It is coarse in texture, works easily, and stands well, but shrinks and checks in drying. The timber is very durable. The trees grow in the region of the Alleghenies from Maine to Michigan and southward to Alabama.

(7) Elm—There are five species of elms found scattered throughout the eastern and Central States. The trees are usually large and of rapid growth, and do not form forests. The timber is hard and tough, frequently, cross-grained, hard to work, and shrinks and checks badly in drying. The wood will take a high polish, and is well adapted to staining. The texture runs from coarse to fine and the color from brown with shades of gray to red.

(8) Gum—The wood of the gum tree is used extensively for cabinet work, furniture, and interior finish. It is of a fine texture and heavy, quite soft, yet strong. It is reddish brown in color. It warps and checks badly and is not durable if exposed. It is also difficult to work. The species of gum used in carpentry is the sweet gum, which grows to medium-sized trees with straight trunks. Though quite abundant east of the Mississippi River it does not tend to form forests.

(9) Maple—Almost all of the maple used in buildings comes from the sugar maple tree, which is most abundant in the region of the Great Lakes, but is also found from Maine to Minnesota and southward to Florida. The trees grow in all sizes and often form quite considerable forests. The wood is heavy and strong and of fine texture. It often has a fine wavy grain which gives the effect known as "curly". It is of a creamy white color, shrinks moderately, works easily, and takes a good polish. It is often used for flooring and other inside finish.

(10) Oak—There are about 20 different kinds of oak found in the United States which may be classified under three different heads as the white, red, and live oaks.

The red oak is usually more porous, less durable, and of a coarser texture than the white or live oak. The trees are of medium size and form a large proportion of all the board-leaved forests. Live oak was once extensively used but has become scarce and is now expensive. Both the red oak and the white oak are used for inside finish, but they are liable to shrink and crack, and must therefore be first thoroughly seasoned. They are of slightly different color, the white oak having a straw color while the red oak has a reddish tinge. Oak is always better if quarter-sawed, when it shows what is known as the silver grain.

(11) Poplar—This wood is also known as white or tulip wood. There are a number of different varieties growing in various parts of the country. The tree is large and is most common in the basin of the Ohio river. It does not form forests. The wood is light, soft, free from knots, and of fine texture. In color it is white or yellowish white and frequently has a satiny luster. It is often stained to imitate some of the more costly woods, such as cherry. It warps badly if not thoroughly seasoned.

(12) Sycamore—This wood is heavy, hard, strong, of coarse texture and is usually cross-grained. It is hard to work, shrinks, warps, and checks badly. The trees grow rapidly and to large sizes. They are found throughout the eastern part of the United States, but are most common along the Ohio and Mississippi rivers.

(13) Black Walnut—This is a wood which used to be extensively used for interior finish and for the manufacture of furniture. It is heavy, hard, of coarse texture, and of a dark-brown color. Although the wood shrinks somewhat in drying, it works easily, stands well, and takes a beautiful polish. The tree is large and of rapid growth. It was formerly abundant in the Alleghany region, and was found from New England to Texas and from Michigan to Florida. It is now becoming scarcer and is most expensive.

7. MECHANICAL PROPERTIES OF TIMBER.

In speaking of wood we are accustomed to use certain words to express our idea of its mechanical properties or of its probable behavior under certain conditions. Thus we say that a wood is hard, tough, brittle, flexible, etc.

a. Hardness.

If a block of wood is struck with a hammer, the resulting impression or dent will be deep or shallow according as the wood is soft or hard. A wood is said to be very hard when it requires a pressure of about 3,000 pounds per square inch to make a "dent" one-twentieth of an inch deep. A hard wood requires only about 2,500 pounds to produce the same effect. Fairly hard wood will be indented by a pressure of 1,500 pounds and soft wood with much less. Maple, oak, elm, and hickory are very hard; ash, cherry, birch, and walnut are hard; the best qualities of pines and spruce are fairly hard; and hemlock, poplar, redwood, and butternut are soft.

b. Toughness.

Toughness is a word which is often used in connection with timber, and implies both strength and pliability such as is found in the wood of the elm and hickory. Such timber will withstand the effects of jars and shocks which would cause woods such as the pines to be shattered.

c. Flexibility.

Timber is said to be flexible when it bends without breaking. A flexible wood is the opposite to one which is brittle. The harder woods, taken from the broad-leaved trees, are usually more flexible than the softer wood taken from the cone-bearing trees. The wood of the main trunk is more flexible than that of the limbs and branches, and moist timber is more flexible than dry wood. Hickory is one of the most flexible of all woods.

d. Cleavage.

Most woods split very easily along the grain, especially where the arrangement of the fibers is simple, as is the case with most coniferous woods. In splitting with an ax, the ax heads acts as a wedge and forces the fibers apart, and usually the split will run along some distance ahead of the ax. The hard woods do not split as readily as the soft woods, nor seasoned wood as easily as green wood.

8. SUITABILITY OF DIFFERENT VARIETIES OF TIMBER FOR CERTAIN PURPOSES.

a. To withstand contact with earth—chestnut, white cedar, cypress, redwood and locust are very good.

b. Light Framing—Timber for this purpose should be free from structural defect such as knots and shakes, and if possible should be obtained in fairly long straight pieces. Spruce, yellow pine, white pine and hemlock satisfy these requirements fairly well; spruce being perhaps a little better than the others.

c. Heavy Framing—A strong timber which can be obtained in large long pieces is necessary for this work. Georgia pine, Oregon pine and white oak may all be used, as well as Norway and Canadian red pine.

d. Outside Finish—A wood which can be easily worked and which will also withstand the effects of the weather is required for all outside finish, such as clapboards, shingles, siding, etc. White pine, cypress or redwood are very satisfactory. All of them may be used for either shingles, clapboards or siding, with the addition of cedar for shingles, and sometimes Oregon pine and spruce for siding.

e. Interior Finish—In general a wood should be used which has a pleasing appearance, while for floors hardness and resistance to wear are additional requirements. Oak, hard pine and maple are satisfactory for flooring, while white pine, red wood, cypress and any of the hard woods such as ash, butternut and cherry are more adapted to the other uses for which interior finish is desired.

FASTENINGS

1. NAILS.

Nails are sold in quantity by the keg, which holds 100 pounds. Common nails are thick and have large flat heads. They are used in rough work where strength is required. Other kinds of nails, differing from common nails in the shape of the head or diameter of the shank and especially adapted to some particular use, are also used. Among them are: flooring, casing, finishing and box nails.

The size, length and diameter in inches of common nails and the number to the pound is given in the following table:

Size	Length	Diam.	Steel Wire G	No. to the lb.
2d	1	.072	15	900
3d	1¼	.08	14	615
4d	1½	.098	12½	322
5d	1¾	.098	12½	254
6d	2	.113	11½	200
7d	2¼	.113	11½	154
8d	2½	.131	10¼	106
9d	2¾	.131	10¼	85
10d	3	.148	9	74
12d	3¼	.148	9	57
16d	3½	.162	8	46
20d	4	.192	6	29
30d	4½	.207	5	23
40d	5	.225	4	17
50d	5½	.243	3	14
60d	6¼	.262	2	11

2. SCREWS.

Screws are manufactured in a great variety of kinds and sizes. The flat head bright wood screw is the only one that will be considered in this manual. Screws are sold in quantity by the box which holds one gross. The size of a screw is designated by stating its length and giving a number indicative of its diameter beneath the head.

3. BOLTS.

The carpenter uses a number of different kinds of bolts, among them machine, carriage, tire, stove, and expansion bolts. The size of a bolt is given by stating its diameter and length in inches.

The machine bolt has a square head.

The carriage bolt has a round head and square shank or body immediately beneath it.

The tire bolt has a countersunk head.

Stove bolts are made either with a flat countersunk screw-slotted head or with a screw-slotted round head.

An expansion bolt is a machine bolt with a special nut so constructed that it expands as the bolt enters it. It is used by first boring a hole of a slightly larger diameter than the nut and a little deeper than its length. The nut is then placed at the bottom of the hole, the bolt inserted and turned home.

The lag screw is a bolt with a square head with a gimlet point and a coarse screw thread from the point toward the head.

ROOF FRAMING

Figure I shows a diagram of the method used to determine the relative lengths of the common and hip rafters. For a roof of 1/3 pitch and unit size (unit size means 1 ft. run) line ac shows the unit length of a common rafter; the length of line ac is 14.42 inches. (See end view Fig. 1). To find the length of the common rafter for any roof having 1/3 pitch multiply the length of run in feet by 14.42 inches. Example: For a roof of 12.5 ft. run, 1/3 pitch, 12.5 x 14.42″=180″, about 15 ft. The unit length of a hip rafter is determined by construction as follows: With a as center (see top view, Fig. 1), and ac as radius, draw arc cc', project point c' to side view and get point c^2. Connect c^2 with a. Line ac^2 is the unit length of hip rafter for a 1/3 pitch roof. It is 18.75″ long. To find the length of hip rafter for any roof of 1/3 pitch, multiply the run by 18.75″.

Figure II, is a pictorial representation of Figure I.

Figure III and Figure IV, show method of making heel and plumb cuts of hip and common rafters.

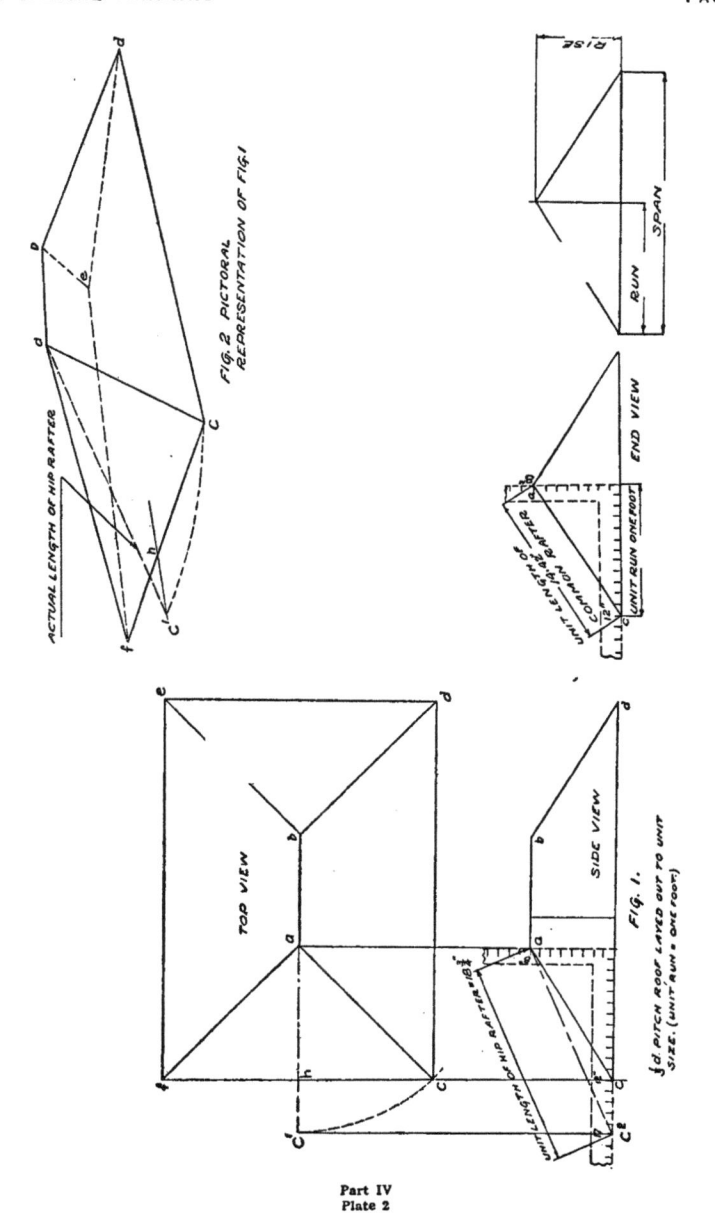

WAR DEPARTMENT
COMMITTEE ON EDUCATION
AND SPECIAL TRAINING

MANUAL No. 4
PART IV
PAGE 13

Part IV
Plate 2

TO LAY OFF THE SIDE CUT FOR HIP RAFTER.

Lay framing square along top edge of the rafter as shown in Figure III (top view), taking 17″ on the tongue and the unit length of the hip rafter for the pitch required as (18.75 for 1/3 pitch) on the blade and scribe along the blade.

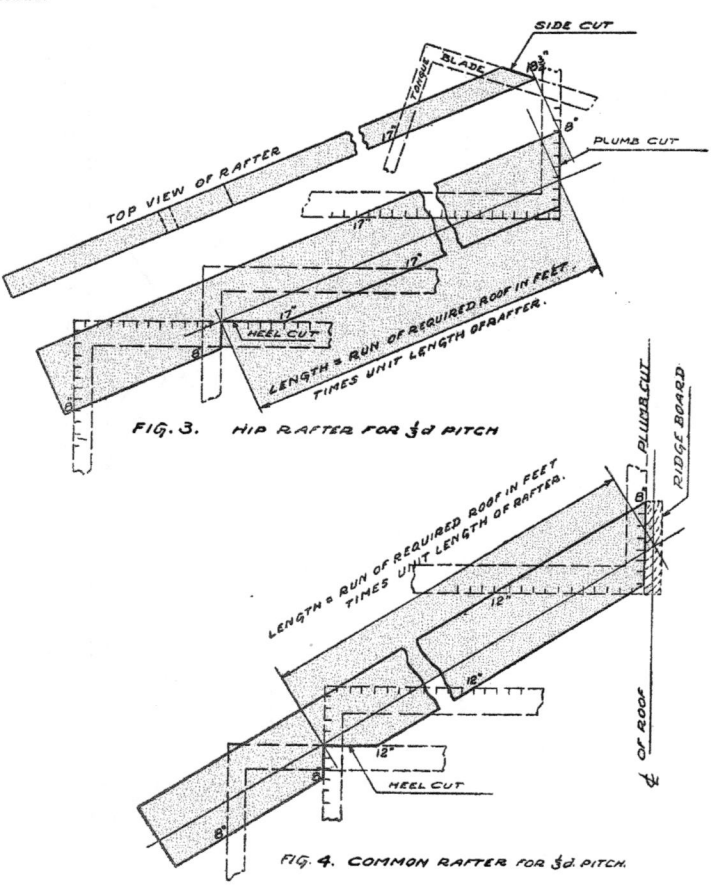

FIG. 3. HIP RAFTER FOR ⅓ PITCH

FIG. 4. COMMON RAFTER FOR ⅓ PITCH

Part IV. Plate 3

WAR DEPARTMENT
COMMITTEE ON EDUCATION
AND SPECIAL TRAINING

STAIR BUILDING

Stairs are made up of three members—the stringers, the risers, and threads. There is one less thread than riser in a stairs. The "run" is the horizontal distance covered by the stairs; the "rise," the vertical distance as from ground level to porch level. A certain relation exists between the tread and riser of a stair, that should be adhered to as closely as possible. This rule is twice the rise plus the tread should equal 24". As an example in stair construction let us take a stairs covering, 3' rise and 3'-6" run, with 5 risers—3' rise=36"÷5= 7 1/5" width of riser; 3'—6" run=42"÷4=10½" width of tread. Therefore 7 1/5" on the tongue and 10½" on the blade of square will give cut in string for tread and rise.

DOOR AND WINDOW FRAMING

Rule for finding size of opening in frame work for windows

To get the vertical measurement between stool and header add 11" to total glass measurement.

The width between studs is gotten by adding 10" to the width of glass measurement.

Rule for finding size of opening in frame work for doors:

To get the vertical measurement from top of joists, add 5" to height of door if double floor is used.

To get the width of opening, add 7" to width of door.

Working drawing
Part IV. Plate 4

WAR DEPARTMENT
COMMITTEE ON EDUCATION
AND SPECIAL TRAINING

MANUAL No. 4
PART IV

NUMERICAL LIST OF JOB SHEETS

Job No. 1 CROSS CUT SAWING. Operations: Lining (pencil and square), sawing, testing with steel square.

Job No. 2 RIP SAWING. Operations: Chalking a line, striking a chalk line, rip sawing.

Job No. 3 CHOPPING TO A LINE (hand ax or hatchet). Operations: Measuring, scoring, paring.

Job No. 4 JOINTING AN EDGE. Operations: Planing, testing with a try square, sighting.

Job No. 5 SAW HORSE. Operations: Dimensioning, drawing, laying out, assembling, (nailing), leveling.

Job. No. 6 NAIL BOX. Operations: Laying out, jointing, shaping with draw shave.

Job No. 7 PLUMB RULE. Operations: Laying out, jointing, parallel edges, use of brace, bit and marking gauge, testing.

Job No. 8 STRAIGHT EDGE. Operations: Laying out, chiseling, testing.

Job No. 9 MITRE BOX. Operations: Boring with gimlet bit and rose countersink, screws and screw drivers.

Job No. 10 STAIR STRINGERS. Operations: Estimating, laying out with steel square or pitch board.

Job No. 11 STEPS. Operations: Estimating, laying out, use of mitre box.

Job No. 12 WORK BENCH. Operations: Estimating, laying out, assembling.

Job No. 13 BUILDING FRAMING. Operations: Estimating, laying out, assembling.

Job No. 14 ROOF FRAMING. Operations: Estimating, laying out with steel square, assembling.

Job No. 15 WINDOW FRAME. Operations: Estimating, laying out, jointing, assembling.

Job No. 16 DOOR FRAME. Operations: Estimating, laying out, jointing, assembling.

TITLES OF UNPRINTED JOB SHEETS*

WIRE ENTANGLEMENT FRAMES. Operations: Cross lap joint, stretching and fastening barbed wire.

RECTANGULAR WATER TANK. Operations: Housed joints, calking.

SHAFT AND GALLERY FRAMING. Operations: Laying out, jointing ground sill and cap sill by means of mortise and tenon joints.

STANDARD "A" TRENCH FRAME. Operations: Laying out with bevel, assembling.

FRAMED TRESTLE BENT. Operations: Estimating, layout, use of bolts.

KING POST TRUSS. Operations: Estimating, laying out, cutting bridge mortise and tenon joints, assembling.

QUEEN POST BRIDGE TRUSS. Operations: Estimating, laying out, cutting bridge mortise and tenon joints, assembling.

BRIDGE (not more than 25 foot span). Operations: Estimating, laying out, bridge joints, splices for tension, use of fish plates, erecting.

---

*At the signing of the armistice, job sheets for distinctly military type of construction problems were in preparation. The above titles suggest the character and scope of the uncompleted section of this manual.